MOOSE

Published by Smart Apple Media
1980 Lookout Drive, North Mankato, Minnesota 56003

Design and Production by The Design Lab/Kathy Petelinsek

Photographs by Craig & Robin Brandt, Tom Stack & Associates, Wide Angle Productions

Copyright © 2001 Smart Apple Media.
International copyrights reserved in all countries.
No part of this book may be reproduced in any form without
written permission from the publisher.

Library of Congress Cataloging-in-Publication Data
Wrobel, Scott.
Moose / by Scott Wrobel
p. cm. – (Northern Trek)
Includes resources, glossary, and index
Summary: Discusses the history of the North American moose, its physical
characteristics and habits, and efforts to preserve the species.
ISBN 1-58340-034-6
1. Moose–Juvenile literature. [1. Moose.] I. Title. II. Series: Northern Trek (Mankato, Minn.)

QL737.U55Q77 2000
599.65'7–dc21 99-29936

First Edition

2 4 6 8 9 7 5 3 1

NORTHERN TREK

MOOSE

WRITTEN BY SCOTT WROBEL
PHOTOGRAPHS BY WIDE ANGLE PRODUCTIONS

SMART APPLE MEDIA

The moose is one of the great giants of North America. With a towering build, massive antlers, and a booming voice, the moose is an impressive animal fearing few enemies in its forest habitat. At the same time, moose are quiet, curious animals that are often willing to live near towns and recreational areas filled with humans. Unfortunately, encounters between moose and humans sometimes have negative results, which is one reason that moose habitats were once in serious danger. Because we better understand moose today, these animals have a healthy population that is spread across the lakes and dense forests of numerous northern wilderness areas.

NORTHERN TREK

NORTH AMERICAN moose (*Alces americana*) are the largest members of the deer family in the world. A bull moose can stand seven and a half feet (2.3 m) tall at the shoulder and may measure 10 feet (3 m) in length. An average bull weighs 1,300 pounds (590 kg), or about as much as a small automobile. Some larger bulls are as heavy as 1,800 pounds (825 kg). Females are typically much smaller than males.

The moose's front end is enormous, with a muscular chest, tall shoulders, and a thick neck leading to a long skull. A flap of skin, called the bell, hangs from the moose's throat. The back legs are slightly shorter than the front ones, making the **hindquarters** much narrower than the front. Although many people think of moose as clumsy, lumbering creatures, they are actually strong, **agile**, and fast. A healthy moose can outrun most **predators** at speeds of up to 35 miles (56 km) per hour.

The hulking brown-coated body of the moose is similar to that of its North American neighbor, the bison.

The moose's nose and soft upper lip hang over its lower lip, allowing the animal to carefully pick leaves, twigs, and berries off bushes and trees. Moose have a fine sense of smell, which makes up for poor vision. The ears are **independent** of each other and can rotate in completely different directions to pick up sounds from all around.

A bull moose's most distinctive feature is its antlers. In older, or mature, bulls, the antler rack can be as wide as six feet (1.8 m) and weigh up to 90 pounds (41 kg). In the spring of a bull's first year, however, the antlers are nothing more than two small, velvet-covered knobs. As the antlers grow and harden, the velvet peels off. Unlike elk

A moose's antler rack takes just four months to grow. The velvet is made up of blood vessels needed for the antlers to develop.

The loud bellowing sound that moose make is called bugling. A moose may bugle to contact a mate or to proclaim its territory, though a moose will not fight to defend its area from another moose that may wander too close.

and deer, moose do not shed their antlers. They keep them for life and use them to forage for food in snow, to fend off predators, and to use as weapons in the fight for mates.

Bull moose are not herd animals, like cattle or buffalo. They usually travel alone when foraging for food. A single moose needs three to four square miles (7.7-10.2 km^2) of land for foraging. Twigs, tree bark, leaves, and plants, including many kinds of water plants, provide all the food a moose needs

The legs of an adult moose are usually 36 to 40 inches (91-103 cm) long, allowing the animal to travel through deep water or snow.

Moose are herbivores, meaning they eat only plants, twigs, bark, and leaves. White birch and aspen trees are important food sources, as they are plentiful even in the harshest winters.

to survive. An adult moose eats 40 to 60 pounds (18-27 kg) of food each day. Moose have been known to stand upright against trees, clawing the trunk with their hooves to loosen bark. To eat lake weeds and lily pads, a moose will dunk its head deep into the water. Moose are strong swimmers and have been known to swim across lakes 15 miles (24 km) wide.

A bull moose will forage alone, while cows and calves group together. Grazing cows and calves are always watching for predators. Their heads bob up after each nibble of grass, and their ears rotate to pick up sounds of danger. The wolf is a natural predator of the moose, but wolves have been nearly wiped out in many moose **habitats**. Bears and humans still threaten moose, however. Cars and trains are also dangers to moose. In Alaska, between 500 and 1,000 moose are killed each year by motorized vehicles.

When **breeding** season begins in late July, bull moose join the herds to fight for the cows. When bulls fight, they lower their heads and ram into each other until their antlers lock together. Then each bull tries to push the other backward. Only the strongest males with the largest antlers may earn the right to mate.

It takes about eight months for a baby moose to develop inside its mother. Newborn calves may

---

Newborn moose have reddish coats that darken as they age. To communicate, they make low grunting sounds, called bleating.

weigh 25 to 30 pounds (11–14 kg). Mothers give birth in **secluded** areas, usually near a river or lake, and remain there with the calf until it is strong enough to move into the forest. A calf can run and swim after only a few days, but it will stay near its mother until the following year, when it's time for another calf to be born.

Moose probably first entered North America 12,000 to 14,000 years ago, when they crossed the **Bering Land Bridge** connecting what is now Alaska and Russia. As they moved south into Canada and

Moose need lots of area. If too many are crowded with other plant-eating animals (such as deer), the moose will get sick and die.

A mature moose can consume between 40 and 50 pounds (18–23 kg) of food each day during the winter, and about 10 more pounds (4.5 kg) of food per day in the summer.

the northern United States, moose lived for thousands of years without ever seeing humans.

Many Native American traditions honor the moose. These animals provided meat, hides for clothes, and bones for tools. When early European explorers arrived in America, the northern forests of the United States, as well as vast stretches of central and western Canada, were full of moose. Before the expansion of cities and farms destroyed much of their natural habitat, moose also lived as far south as Virginia.

By 1880, moose had become a threatened species in the United States. Even in the thick forests of northern Minnesota and Michigan, overhunting and expanding human settlements drastically reduced moose populations. In the early part

of the 20th century, efforts were first made to protect the animal. Moose hunting was outlawed in many states, and large sections of the animal's natural habitat were protected by law. By the 1990s, moose populations in Canada, Alaska, and the northern United States were thriving once again.

Today, nearly one million moose live across North America, with some ranging as far south as parts of Massachusetts. As long as its habitat is not destroyed by human development, the towering moose can continue to roam the great northern woods for generations to come.

To protect moose in the future, scientists have been studying vast areas of land to understand how populations of moose and their natural enemies—wolves and bears—coexist year after year.

VIEWING AREAS

HERDS OF MOOSE COWS

and calves may be easier to spot than males, which roam alone in the dense forests of the north. The best viewing areas are national and state parks in the northern parts of the United States.

With the continued support of people and the government, the moose population will continue to grow. Listed here are various moose habitats with public access. As with any trek into nature, it is important to remember that wild animals are unpredictable and can be dangerous if approached. Bull moose are especially aggressive during mating season and should not be disturbed. The best way to view wildlife is from a respectful—and safe—distance.

BERING LAND BRIDGE NATURAL PRESERVE IN ALASKA

The numbers of moose are constantly increasing in the western part of this massive, remote preserve in northern Alaska.

BOUNDARY WATERS CANOE WILDERNESS AREA IN ELY, MINNESOTA

Motorized vehicles are not allowed in the wilderness area, but you can enter the lake-filled forest by path or by canoe. There is a great chance of seeing many moose foraging in the lakes and streams, just yards from your canoe. Also, many moose are seen foraging along the edges of the BWCWA, where logging has occurred.

ISLE ROYALE NATIONAL PARK IN HOUGHTON, MICHIGAN

Your chances of seeing a moose close-up in a deep woods setting are probably greatest here. The moose population is very high for the amount of habitable space. You may have to rest on your hike as a moose or two feeds near your trail.

JASPER NATIONAL PARK IN ALBERTA, CANADA

Visitors can often see many moose as they drive along the roads in this large park.

Glossary & Index

agile: *able to move quickly and easily*

Bering Land Bridge: *a strip of land that rose above sea level thousands of years ago connecting the Asian and North American continents*

breeding: *mating between a male and female animal to produce offspring*

habitat: *a place or environment where a plant or animal normally lives*

hindquarters: *the back legs and rear end of an animal*

independent: *something that works by itself without helping or relying on anything else*

predators: *animals that kill other animals for food*

secluded: *separate and hidden from view*

antlers, 9–10
breeding, 15
bugling, 9
calves, 15, 17
conservation, 18
ears, 9
eyes, 9
feeding, 9, 10, 15, 18
habitat, 10, 17–18
history, 4, 17
hunting, 19
laws, 19
mouth, 9
Native Americans, 18
population, 19
predators, 15
size, 6
swimming, 15, 17
viewing areas, 23